With best wishes of
Fred. E. Jacob
Major AUS. AF. Ret

# Takeoffs and Touchdowns

# TAKEOFFS AND TOUCHDOWNS

## My Sixty Years of Flying

Fred E. Jacob

SAN DIEGO
A. S. BARNES & COMPANY, INC.
IN LONDON:
THE TANTIVY PRESS

First Edition
Manufactured in the United States of America

For information write to:
A.S. Barnes & Company, Inc.
P.O. Box 3051
La Jolla, California 92038

The Tantivy Press
Magdalen House
136–148 Tooley Street
London, SE1 2TT, England

Library of Congress Cataloging in Publication Data

Jacob, Fred E.
Takeoffs and touchdowns.

1. Jacob, Fred E. 2. Air pilots—United States—Biography. I. Title.
TL540.J26A37 629.13'092'4 [B] 80-28865
ISBN 0-498-02540-3

2 3 4 5 6 7 8 9 84 83 82 81

To Royal; to the General; and to Evelyn, my Queen.

# Contents

# Acknowledgments

The writing of a book usually leaves the author indebted to many people. Seldom can one person be the only factor, or even more than a major influence, in developing an idea into a manuscript. Consequently, I owe first an expression of deep appreciation to my one and only, Evelyn Annie Jacob. After 58 years of marriage, and 53 years with her beside me in the copilot's seat of many airplanes, she is obviously the inspiration and critic of every phase of this book and my life in general. She, and many others, contributed to the labor of love that produced this book.

The list of teachers and pilot friends who have touched my life and activities was so long that although I gratefully acknowledge their impact, I can cite individually only those who most recently worked directly in the preparation of this work.

Our friend of many decades, Lt. Gen. Jim Doolittle, took time out of his busy schedule to read every page of the manuscript and make some excellent additions and comments.

Perhaps the most unsung member of any writing team is the secretary-typist. In this case it is most certainly true. Our loyal Rhoda Lukasiak, with us many years in the business world and now in the writing field, has had to read and decipher both tapes and handwriting till her eyes must have ached. But never a kick, just a smile, and always: "When will you have that next chapter ready for me?" That push kept us going.

Art Seidenbaum, book editor of the *Los Angeles Times* deserves a gold star after his name for referring us to my publishers, Cynthia and Charles Tillinghast. They consented to take a look at the manuscript and decided to publish it. But the most satisfying piece of luck from this author's standpoint was their assignment of one Mike Werthman, their senior editor, to handle the project from start to the finished book. His attention to detail, catching errors of dates, places, airplane numbers, etc. has been fantastic. His contributions to the whole work are such that the author feels he should almost be considered a co-author. If this is readable and free flowing it is Mike's deft touch that has made it so, for which we are most grateful.

If you, the reader, have had the thrill of operating the controls of an airplane, there are incidents in our experiences that will bring back memories. If as yet you have not been in the pilot's seat, may we but hope that our sixty years of experiences will cause you to start planning toward that end.

# Preface

To an earth-bound human, the sweetest sound this side of heaven may be the music of a famous orchestra. Or it can be the deep bass tones of an organ playing a Bach concerto. Or perhaps a band giving forth with a rousing rendition of Sousa's "Stars & Stripes Forever." But to a pilot, there's one sound that is much more soothing and satisfying—the smooth steady M-M-M-M-M of the engine putting out its hum of perfectly tuned explosions. This is true whether it be a 65-horse corn popper on a little Cub or four fans in perfect "sync" pulling you along with thousands of horses. Today's kerosene jet blast furnaces don't make an equally satisfying roar.

Thus it was one pre-dawn morning in 1942 when five of us were crewing our B-24 bomber at 10,000 feet as we passed just east of the mouth of the Amazon River in Brazil. The very tops of the high cumulus off to our right were getting a rose petal pink but to us it was still dark purple. Off to our left was where old Sol would soon make his appearance out of the South Atlantic. Then it happened: a first little spear of light hit us and the clouds off to our right came alive with color as we passed by like tourists—at our 160 mph cruise speed.

Hans Lyon, our trusty engineer sergeant, a native of New Jersey, was standing behind me in my left hand pilot's seat.

"Heaven!" he commented mostly to himself. "Bout as near as I'll ever come to seeing it," he added. Copilot Jerry in the right hand seat nodded. As for me, I was doing nothing, with the auto pilot doing the work, I was free to take in the sight and file it away as a memory of a ferrying trip, delivering a bomber to the fighting crews in Africa and east.

Sergeant Lyon, knowing my weakness for pictures, reached down and handed me my 16mm movie camera from the musette bag behind my seat and I took a series of shots of those clouds and the passing Amazon River through the two props on the right side of the plane. Of all the thousands of feet I shot during my World War II tour of duty, this footage best proves that a pilot is indeed a favored human, seeing unforgettable sights and listening to the music of perfectly tuned engines up front.

In the life of any pilot with decades of flying behind him, you can be sure of two things. First, that he (or she) will have several fat logbooks full of memorable flights. Second, that he or she will have a fair percentage of hearing loss from the extra high decibels put out by the short stack engines of old, the Hispano-Suizas, the OX-5s, the big rotary R-2800s, and others that followed.

As one who qualifies on both the above counts, starting with my first sighting of an "aeroplane" in 1910 through more than sixty years of active participation, I continue to feel that this fabulous flying game is still in its infancy. I hope that by sharing with you my experiences of those six decades it will bring back memories of your own if you are a pilot.

If you are not yet a licensed pilot, I'd like to think the lessons I learned and the good times I enjoyed will strengthen your resolve to go ahead with the labors of getting that license and experience. You will need it to keep up with the younger generation as you live toward the glorious years of the coming twenty-first century.

I've tried to make this voyage of recollection both pleasurable and memorable. And if it inspires you to become a more active pilot or copilot of a family plane, I'll be more than satisfied.

Fred E. Jacob
Glendale, California
February 1981

# Introductory Comments
# by
# Lt. Gen. James H. Doolittle

I read *Takeoffs and Touchdowns*, by Fred E. Jacob, with amusement and interest. It is the story, enlightening and entertaining, by a perceptive aviator, of his flying experiences in the early and intermediate days of aviation.

Particularly enjoyed the saga of the barnstorming era of the 20s and 30s. The aviation fraternity was small—and highly migratory—so everyone knew practically everyone else. It was a treat to "visit with" old friends: Slim Lindbergh, Phil Love, Les Arnold, C. C. Moseley, Joe Plosser et al.

During the great depression the aviation industry, one of the newest, was hardest hit. A pilot, if asked what was the greatest hazard in aviation, was apt to answer, "starvation."

I was also intensely interested in the World War II section which is very well done. Again it was possible to visit with old friends and relive experiences. Events and the author's personal experiences were depicted clearly and convincingly.

Carmel, California
October 14, 1980

## THE LAST FLIGHT

When the last long flight is over
And the happy landing's past
When the altimeter tells me
That the crackup's come at last
I will swing her nose for the ceiling
And I will give my crate the gun
I'll open her up and let her zoom
For that blinding airport of the sun.
Then the great God of flying men
Will smile at me, sort of slow
As I stow my ship in the hangar
On the field where flyers go.
And I'll look then upon his face,
The Almighty, flying Boss,
Whose wingspread fills the Heavens
From Orion up to the Cross.

BILLIE.

Pilot who wrote this, killed by crashing at Lick Obervatory, Mt Hamilton.
May 21 1939 — 7

# Takeoffs and Touchdowns

The Jacob farm in central Michigan, where I arrived in 1899. Photo taken by me in 1913 or 1914, with a glass-plate 5x7 view camera on a tripod. I even developed it in the basement with Sears chemicals and instructions.

# 1

# Let's Begin at the Beginning—I Did

### *The Great Day In June 1910*

Central Michigan in June is hot! And with lakes on three sides it also is humid, as it was on a farm north of Lansing. The wheat harvest was in full swing. The binder, with two horses pulling it, was throwing a steady flow of wheat bundles. Two wagons were loading them to be hauled to the barn. In those days everybody worked, especially boys such as my brother and I, ages 9 and 12 respectively. Each of us was driving a team on a wagon and the hired men were loading the bundles as we drove forward doing our part of the job. Now and then Shep, the family dog, would kick up a rabbit and have a little run just for sport. Occasionally, he would catch a woodchuck on a safari too far from his hole and would have a peck of fun for a half hour capering around said animal until he finally decided to dispatch it.

All at once Shep started a furious barking. "Huh? Cornered another woodchuck," said the hired man loading bundles on the wagon. But Shep didn't move. His nose was pointed west and looking up, not down at the ground. Just then the horses started prancing with little mincing steps. This caused the hired man on the wagon to grab the reins away from me before things got out of hand, which they almost did. By that time we could hear a roar far in the distance—but still a roar. Funny, too. Bright sunshine, no clouds, no thunderstorm, and no sign of a "twister." So what could it be?

The horses continued to prance nervously and Shep continued to bark. A whoop from the man on the other wagon: "AEROPLANE!" By that time the teams were standing on their hind legs thoroughly panicked by this tremendous noise from the approaching thing.

Somehow or other I, a twelve-year-old boy, was instantly on the ground jumping up and down screaming as the aeroplane approached directly toward us some 500 or 600 feet overhead. It was making a horrendous noise from the four cylinders without mufflers. The pilot sitting out in front was clearly visible. In another thirty seconds he was overhead. Soon, he was fading in the distance at a good 45 mph.

Everything stopped. The event was discussed by my father, John, and the assembled workmen. It was the event of the week, the month, yes, and of the year for us. The country paper had mentioned the state fair in Grand Rapids and another in Detroit and we figured that this was the Curtis Pusher airplane, piloted by a local celebrity, Mr. Parmalee, enroute from the Grand Rapids to the Detroit fair. We had had ringside seats to this, the first airplane that any of us had ever seen.

At lunch that noon, as the event was recounted to the women folks, from the far end of the table I made a solemn announcement: "I'm going to fly one of those things some day!"

That was a statement that I had reason both to regret and to recall fondly many times. And right after I made it I began to be teased by the hired men, who reminded me that I couldn't stand climbing higher than some thirty feet—"And you want to fly aeroplanes?"

### *School Days*

A small fraction of the eighth-grade graduates made it into high school in those days. Father was opposed to it, saying it made loafers and bums out of otherwise good men when they got high school and especially college diplomas. Mother plus aunts and uncles overruled him and off I went to high school in the Fall of 1912.

In high school I read everything about flying in *Popular Mechanics* and other magazines in the school library. I still was convinced I would someday fly, despite my continued fear of heights.

I was kept fully occupied in high school with football, baseball, band, and the college preparation course to which my father had

**Fred the first baseman for St. Johns High in 1916.**

grudgingly agreed and which gained me admission to the University of Michigan in 1917.

Without a doubt the biggest event of my Freshman year at Ann Arbor was the arrival of two airplanes from Selfridge Field some 45 miles to the north. Fortunately, these two pilots decided to land in a meadow east of town owned by a cousin of mine, Floyd Uloth. They made a bargain with him to back their planes into his toolshed while they were on campus taking some courses in map reading. Naturally, my bicycle and I were out there on the double, arriving by the time they had finished negotiating with Cousin Floyd. Captain Smith (whom I later met when he was a colonel) took a liking to me and I got to ride with him several times as he took his plane up for short trips. He left the dual controls in and I could see what made

**In 1918 two planes came down from Selfridge Field, some 45 miles from Ann Arbor, Michigan. I got my first dual in one of them, a Gates Standard. While coming into my cousin's pasture, one of the ships got caught in a sudden crosswind. Result—torn up fabric and ribs, but quickly repaired by Selfridge mechanics and back flying.**

the thing go up and down and how the rudder pedals made it go sideways or into turns. In all, I got in probably two to three hours of dual merely following him through very lightly.

Their course ended and he gave me my last ride one nice Saturday morning. Both planes were in the air. I was riding in the one not shown in the picture. The young lieutenant who was flying the ship shown got hit by a strong gust of crosswind while landing. One wing brushed an apple tree. Though he completed the landing, after a fashion, some parts were cracked. Down came the repair crews from Selfridge and had it flying in a day or two. I am quite certain their patching of spar and ribs wouldn't pass the regulations of today, but it flew OK.

However, a black cloud was developing in the sky for me. They were taking enlistments in Ann Arbor, Detroit, and other cities for the Signal Corps. I was at the head of the line. A paunchy little doctor with his stethoscope checked me over and that was OK. He had me dance on first one foot and then the other and said, "Hold it. Do that again on that left foot." I danced again. There was an audible click as a result of a bone broken in high school football. It didn't bother me but it did click. Result: REJECTION. Two more tries elsewhere and still rejected. I was heartbroken that I couldn't get into the Signal Corps. In my safety deposit box I carefully keep those three rejections from World War I. The reason they rejected me, "You can't march twenty miles a day on that ankle." My reply was, "Who the H- wants to march twenty miles a day!" So, no flying though many of my friends made it and passed into the flying corps of the Army, some later to be killed—as will be discussed subsequently.

### *The Quirks of Fate—Denver, Colorado*

In spring of 1919, I was diagnosed as having tuberculosis. Naturally, I was pretty much downcast but took the advice of one of the doctors to go to Denver, which was then the health capital of the country for TB. Off to Denver with a letter to a Doctor Lockhard, a former Michigan professor who had become a "lunger" himself and had moved there to become an arrested case and set up a very thriving tuberculosis practice.

Dr. Lockhard greeted me with his hand out. "You're Jacob."

"Yes, but how did you know, doctor?"

"Why I read the *Michigan Daily* and you threw the discus down at Illinois two weeks ago—broke the record!"

It turned out that Dr. Lockhard was one of those perennial sophomores who even though long out of school kept in touch with everything that was going on on the campus through the publications. Needless to say, I got fast action from Dr. Lockhard and his associates.

They all agreed. The diagnosis of the Ann Arbor doctors was wrong. I didn't have TB. But they didn't know what was making my neck continue swelling till I couldn't eat anything but soups. Couldn't get my mouth open far enough. Finally, as a last resort they sent me next door to a dentist who managed to take some X-rays by working the films in place with his finger. His findings: Three impacted wisdoms! Naturally, there was a celebration. A three hour and ten minute operation sitting in one chair all the time left me with three big holes and no wisdom teeth. It also left me "stranded" in Denver. Before leaving Ann Arbor, I had arranged to take (and passed) my final examinations. There was no pressing reason to return to Michigan immediately, but I'd need to make a living while I stayed on in Denver.

The best way I could think of to find a job was to hit the various studios and Kodak places. Back when I was ten years old, a

**I was a pretty good discus thrower on the University of Michigan track team, 1918–21.**

neighbor woman had loaned me her little box camera with a rubber sleeve in the side through which you changed the 3″×3″ glass plates. With that start I had worked up to two cameras. While at the University I developed films and used a 5×7 Press Graflex for a local photo shop. I even turned out pretty good pictures for the *Michigan Daily* and *Detroit News*.

The manager at Eastman Kodak examined me quizzically, fired a few questions, and took me upstairs in the adjoining building where they had big facilities for developing up to 1,000 rolls of film a day for hordes of tourists who came to Denver in the summer those days.

Unknown to me, my predecessor had been a chap whose work in the darkroom was OK. However, when the girls who sorted and wrapped the prints had to come with questions, he had wandering hands that caused no end of trouble. As a result, one of them, a Miss Evelyn Austin, had given the boss strict orders that he was to hire a woman to develop films, as they didn't want another man in there giving them a hard time. He showed up with me, a man, and introduced me to the various girls and Miss Austin. She gave me a very icy, "Pleased to meet you, Mr. Jacob." Obviously, she wasn't pleased.

Funny thing, though, in a couple of weeks I found that she was a college girl home from Colorado College for the summer, much different from the other girls. Soon I was taking her out to lunch. Then I was out at her home. Not long after the love bug bit for the first time in my life. My return to Ann Arbor that fall confounded the doctors who had written me off as a one way ticket to Colorado. I played on the football and track teams again. Played in the band and was in better shape than ever, now that I had an interest in Denver. We kept the mails busy.

The summer of 1920 I was back in Denver, this time armed with a diamond ring. It took a bit of selling to convince her that she should wear it, but finally I made the sale. I might add that in a family of English descent—both of her parents had been born in England—it was customary for the young man to confer with the father before presenting a diamond ring. Since I did not ask him for her hand but let her spring the ring on him, it caused a little commotion in the Austin family. However, it stuck and it's still there more than 60 years later as I write this. Graduation for both of us in 1921, marriage in 1923, and deep into the business world from then on was my good fortune.

### *A Sad Flight to Remember in 1921*

Fresh out of the University of Michigan with a journalism degree, I was an eager young newspaper reporter and staff writer on a Scripps-Howard newspaper there in Denver.

The average person doesn't realize how much freedom and respect, yes, and cooperation, a newspaperman usually gets. Particularly from politicians. Thus it was that I, a chap of twenty-two, had the freedom of Mayor Stapleton's office, of the Governor's office, and most anyplace I wanted to go looking for a story.

This particular day I was in the office of Frank Lowry, the head of Streets and Parks in the old City Hall in Denver. I had given Mr. Lowry some nice write-ups with pictures, that did no harm to his political image around the city, so he always had some kind of a story for me. With my scratch pad in hand I was talking to his secretary when he came out and put his arm around my shoulder but with rather sad mien. "I guess you've heard, Jacob, they're bringing the body of my boy home from France in another week or two. We don't know the exact date yet, but I'll give you the story as soon as we do. Maybe if you want to run a little advance notice on it, it wouldn't do any harm."

I started writing and when I took down the name of Frances Brown Lowry something clicked. "Did your son attend the University of Michigan before he enlisted?"

"Why yes, he did. How come you ask that?"

"Because I knew your son when we were both students there four years ago. I did my dawgondest to enlist but with a defective ankle I didn't make it. Your son did. I never knew that he had been a casualty until just now and it shocks me greatly. He was a wonderful chap and a good student; well liked there in Ann Arbor."

That took Mr. Lowry by considerable surprise and my laudatory remarks brought tears to his eyes. By that time we were back in Mr. Lowry's office and there sat a picture of his late son in a frame on his desk.

"Is there any way, Mr. Lowry, that I can get a copy of that picture or borrow that one for an hour or two so we can make cuts from it? You may rest assured that I am going to play this up. It's a real news event." He was very much pleased and that's the way it was handled.

The Governor had ordered a pilot from the National Guard to be ready to fly over the funeral procession from St. Mark's Episcopal Church out to Fairmount Cemetery dropping flowers on the funeral procession. At Mr. Lowry's suggestion I was to go along in the cockpit, also with a basket of flowers. While out at the airport getting our baskets of flowers and checking by phone as to probable time of departure I mentioned to the pilot rather casually but very indefinitely that I had had some stick time back in Ann Arbor. (I didn't tell him that I had had just a couple of hours of dual on a couple of Selfridge Field planes.) His reply was, "Oh, in that case, Jacob, I'll leave the duals in and we'll take turns. When I'm dropping flowers you'll fly and when I'm flying you drop off the flowers." Like a damned fool I agreed.

Everything went off on schedule as we headed east down Colfax Avenue and he made the first two passes down to three or

**Just as we were prepared to fly over the funeral of my school chum, Francis Brown Lowry, at Denver in 1921. Had a close call, thanks to baskets of flowers that got jammed into Captain Hill's stomach when I came back on the stick to pull out of a glide over the funeral procession.**

four hundred feet above the procession and I dropped down handfuls of flowers that may or may not have fallen on the funeral procession. Then it was my turn and I lined up as I had watched him do and gradually lost altitude with the throttle pretty well back. At about the three-hundred-foot height I started coming back on the stick but it wouldn't come back. I knew it had to come back so I really hoisted it back hard. The pilot let out a yell and I brought it back some more. We came down probably a hundred feet above the procession and with the noise of that Hisso engine I am sure we upset things considerably on the ground. After a bit we began to gain altitude and he shook the stick and took over. We got up a thousand feet and he pulled the throttle back to tell me what had happened. His belly and the peck size basket of flowers lined up directly in front of the stick and when I came back on it he couldn't get the basket out of the way. The harder I came back the more the basket was locked in place. From then on we got our baskets of flowers to one side and made the drops the rest of the way to the cemetary without getting below three hundred feet.

When we got back to the airport and took a look the two control sticks didn't line up. I had bent the fitting on mine or on his so that they were a good fifteen or twenty degrees out of parallel. Luckily neither one of the fittings broke or we might had rolled things up in a ball ahead of the procession. There were some comments from the undertaker later but we didn't bother to explain exactly what happened. Whenever I pass over Lowry Air Force Base and Fairmount Cemetary in Denver, that close shave and the fact that war claimed a very good friend of mine from student days come to mind.

FLY YOUR MAIL
TIME FLIES SAVE TIME
WESTERN AIR EXPRESS (INC)

No. 68

This Certifies that Fred C. Jacob
on August 7 1928, traveled via Western Air Express Airline from Denver to Pueblo
over Contract Air Mail Route No. 12.

Time of Departure 7:05

Time of Arrival 8:30 Total Flying Time

Maximum Altitude 3000 Maximum Speed

Air Mail Cargo lbs.

Total weight carried lbs.

Royal Leonard
Pilot Ship No. 12

**My ticket to Pueblo, Colorado, August 1928.**

# 2

# Royal Leonard
## Pilot, Genius and Friend

I had noticed that Western Air Express was running two flights a day between Cheyenne, Denver, Colorado Springs, and Pueblo, a little 200-mile stub run South from Cheyenne, Wyoming, where it tied in with the main line of United Airlines. Flying one of these flights was a chap by the name of Royal Leonard the day in 1928 when I decided to ride to Pueblo with him. My ticket stub bore number 68, so apparently I was the 68th passenger they had in the three or four months that they had had the contract. Pilot Leonard outfitted me with a pair of coveralls, a parachute, a helmet and goggles, and put me in the front cockpit along with two sacks of mail, each weighing ten or twenty pounds at the most. Off we went in the Stearman biplane that brought back memories of the 1917 planes but was much faster. According to my receipt, it took an hour and twenty-five minutes with a stop in Colorado Springs to drop a ten-pound sack of mail. Leonard, being single and a graduate of Brooks and Kelley Fields in the Army, was naturally up on a pedestal in my estimation. Soon we had him out to the house regularly as I absorbed more information regarding flying from him.

In a matter of months, Royal and I decided to purchase a Lincoln Page plane made in Lincoln, Nebraska. We had catalogs and literature on several makes, Swallow, Travelair, and the locally built Eagle Rock. But Royal, as a grad of Kelley, had contacts with Lindbergh who, of course, had graduated some years before. It was the comments of Lindbergh to Royal that sold us on purchasing a Lincoln Page. He had learned to fly off the Lincoln field about 1923 and said their plane was ruggedly built and fast. The word "fast" appealed to us. So we signed an order—for $3,176.60. What really clinched the deal for Lincoln Page was that they would build it up with a 180-horsepower Hispano-Suiza water-cooled engine instead of the usual OX-5, 90-horsepower engine that was so widely used in those days.

Soon the word came that our plane was ready and back we went to Lincoln, Nebraska. It looked good. We were a couple of happy young lads who planned on making a million, barnstorming Colorado and Wyoming, in Royal's spare time when he wasn't flying

**Off to Pueblo with pilot Royal Leonard. The beginning of a long friendship and a big change in the lives of the Jacob family.**

**Royal Leonard, a captain in the Colorado National Guard, beside one of the Consolidated biplanes that were the mainstay of their fleet.**

the mail. It didn't quite work out that way.

Royal immediately started giving me dual instruction. Of all the planes to take dual on, a Lincoln Page Racer wasn't the one, but it was the only plane we had. This was the day of no brakes, no lights, no battery, no nothing except five basic flying instruments. But she was fast! And tricky.

There at the factory airstrip where Lindbergh had learned to fly, Royal asked for a parachute. That seemed to upset both Ray Page, the President of Lincoln Aircraft, and his factory superintendent, Mr. Sloniger. Royal strapped on the chute, climbed into the rear cockpit, and motioned for one of the staff to swing the prop. Propping an airplane in those days was risky business. Many were

LINCOLN AIRCRAFT COMPANY
2409 "O" STREET
LINCOLN, NEBRASKA

BILL OF SALE

KNOW ALL MEN BY THESE PRESENTS:

That we, the undersigned Lincoln Aircraft Company, Lincoln, Nebraska, for value received, do hereby sell, transfer and convey unto the Denver Aircraft Sales Company, 1310 South Humboldt, Denver, Colorado, the following goods and chattels, to wit:

1 - LP-3 Passenger Airplane
Factory No. 251
Motor - Model "I" Hispano
#49394
Ident. No. 7644

Signed at Lincoln, Lancaster County, Nebraska, this 15th day of September, 1928.

LINCOLN AIRCRAFT COMPANY

By Ray Page
Ray Page, President.

S

All agreements are contingent upon labor trouble, accidents, delay of carriers, or any other circumstances beyond our control. All quotations of prices or changes in models are subject to change without notice or prior sale.

**Ray Page, President of Lincoln Page Aircraft when Royal and I bought our plane.**

killed and maimed in doing it.

The Hisso let go with a mighty roar, as it had the short stacks. Royal taxied out to the end of the grass strip and with a quick burst of power was in the air in almost nothing flat compared to the performance of 90-horsepower OX-5 jobs. I noticed that both Page and his staff seemed very much perturbed over what Leonard was up to. I was too ignorant to understand why.

Leonard climbed to above 4,000 feet, following his training at Brooks and Kelly Fields, and pulled her up into a couple of stalls to test her characteristics. Finally, he kicked her over into a right spin and how she did wind! Not one, or two, but three turns and he brought her out rather hesitantly. Back up to 4,000 and left spins, all